EQUESTRIAN JOURNAL

THIS JOURNAL BELONGS TO:

_____

CONTACT INFORMATION:

_____

_____

GOALS

- ☐ _____
  _____
  _____

- ☐ _____
  _____
  _____

- ☐ _____
  _____
  _____

- ☐ _____
  _____
  _____

PLANS

_____

_____

_____

_____

_____

_____

_____

_____

_____

_____

_____

_____

JANUARY

SUNDAY	MONDAY	TUESDAY	WEDNESDAY
4	5	6	7
11	12	13	14
18	19	20	21
25	26	27	28

2026

THURSDAY	FRIDAY	SATURDAY	NOTES
1	2	3	
8	9	10	
15	16	17	
22	23	24	
29	30	31	

FEBRUARY

SUNDAY	MONDAY	TUESDAY	WEDNESDAY
1	2	3	4
8	9	10	11
15	16	17	18
22	23	24	25

2026

THURSDAY	FRIDAY	SATURDAY	NOTES
5	6	7	
12	13	14	
19	20	21	
26	27	28	

MARCH

SUNDAY	MONDAY	TUESDAY	WEDNESDAY
1	2	3	4
8	9	10	11
15	16	17	18
22	23	24	25
29	30	31	

2026

THURSDAY	FRIDAY	SATURDAY	NOTES
5	6	7	
12	13	14	
19	20	21	
26	27	28	

APRIL

SUNDAY	MONDAY	TUESDAY	WEDNESDAY
			1
5	6	7	8
12	13	14	15
19	20	21	22
26	27	28	29

2026

THURSDAY	FRIDAY	SATURDAY	NOTES
2	3	4	
9	10	11	
16	17	18	
23	24	25	
30			

MAY

SUNDAY	MONDAY	TUESDAY	WEDNESDAY
3	4	5	6
10	11	12	13
17	18	19	20
24	25	26	27

2026

THURSDAY	FRIDAY	SATURDAY	NOTES
	1	2	
7	8	9	
14	15	16	
21	22	23	
28	29	30 / 31	

JUNE

SUNDAY	MONDAY	TUESDAY	WEDNESDAY
	1	2	3
7	8	9	10
14	15	16	17
21	22	23	24
28	29	30	

2026

THURSDAY	FRIDAY	SATURDAY	NOTES
4	5	6	
11	12	13	
18	19	20	
25	26	27	

JULY

SUNDAY	MONDAY	TUESDAY	WEDNESDAY
			1
5	6	7	8
12	13	14	15
19	20	21	22
26	27	28	29

2026

THURSDAY	FRIDAY	SATURDAY	NOTES
2	3	4	
9	10	11	
16	17	18	
23	24	25	
30	31		

AUGUST

SUNDAY	MONDAY	TUESDAY	WEDNESDAY
1 / 2	3	4	5
9	10	11	12
16	17	18	19
23	24	25	26
30	31		

2026

THURSDAY	FRIDAY	SATURDAY	NOTES
6	7	8	
13	14	15	
20	21	22	
27	28	29	

SEPTEMBER

SUNDAY	MONDAY	TUESDAY	WEDNESDAY
		1	2
6	7	8	9
13	14	15	16
20	21	22	23
27	28	29	30

2026

THURSDAY	FRIDAY	SATURDAY	NOTES
3	4	5	
10	11	12	
17	18	19	
24	25	26	

OCTOBER

SUNDAY	MONDAY	TUESDAY	WEDNESDAY
4	5	6	7
11	12	13	14
18	19	20	21
25	26	27	28

2026

THURSDAY	FRIDAY	SATURDAY	NOTES
1	2	3	
8	9	10	
15	16	17	
22	23	24	
29	30	31	

NOVEMBER

SUNDAY	MONDAY	TUESDAY	WEDNESDAY
1	2	3	4
8	9	10	11
15	16	17	18
22	23	24	25
29	30		

2026

THURSDAY	FRIDAY	SATURDAY	NOTES
5	6	7	
12	13	14	
19	20	21	
26	27	28	

DECEMBER

SUNDAY	MONDAY	TUESDAY	WEDNESDAY
		1	2
6	7	8	9
13	14	15	16
20	21	22	23
27	28	29	30

2026

THURSDAY	FRIDAY	SATURDAY	NOTES
3	4	5	
10	11	12	
17	18	19	
24	25	26	
31			

Basic info	_____
	_____
	_____

Condition	_____
	_____
	_____

Quirks	_____
	_____
	_____

| Diet | _____ |
| | _____ |

HORSE

FARRIER _____ **#** _____

SHOE SIZE ^F _____ ^B _____

SHOE STYLE ^F _____ ^B _____

USUAL COST ^$ _____

FARRIER APPOINTMENTS

_____   _____

_____   _____

_____   _____

_____   _____

NOTES

Medical

DATE	VACCINATION	GIVEN BY	NEXT DOSE DUE	PRICE

NOTES

HORSE

DENTAL

PERFORMED BY _____

NEXT DUE _____

COST $_____

MEDICATIONS

MEDICATION	INDICATION	DOSAGE/ FREQUENCY	PRICE

NOTES

CARE

BodyWork

DATE	THERAPY TYPE	GIVEN BY	NEXT DUE	PRICE

HORSE

DEWORMER & SAND RID

MEDICATION	DOSAGE/ FREQUENCY	NEXT DUE	PRICE

NOTES

CARE

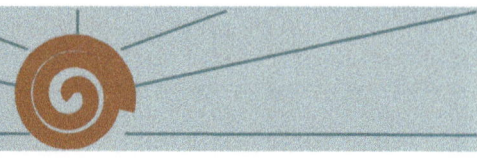

TACKROOM WISHLIST

ITEM- WHAT I NEED, WHOSE IT FOR	PRICE

TACKROOM

EQUIPMENT LOANED OUT

ITEM	WHO HAS IT	CONTACT

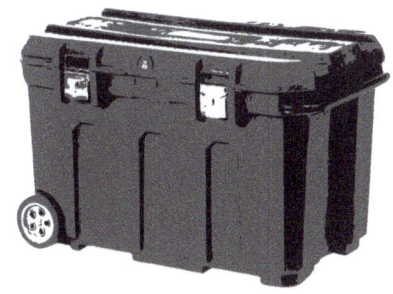

MANAGEMENT

DAY

TIME

ARENA/ LOCATION

LUNGING _____ W T C

WARM-UP _____ W T C

PEAK RIDE _____ W T C

COOL DOWN _____ W T C

MOOD

INSTRUCTOR _____

BEFORE RIDE   DURING RIDE

TAKEAWAYS

_____

_____

IDEAS FOR NEXT RIDE

DAY

TIME

ARENA/ LOCATION

LUNGING _____ W   T   C _____

WARM-UP _____ W   T   C _____

PEAK RIDE _____ W   T   C _____

COOL DOWN _____ W   T   C _____

MOOD

INSTRUCTOR _____

BEFORE RIDE   DURING RIDE

TAKEAWAYS

_____

_____

IDEAS FOR NEXT RIDE

DAY	TIME	ARENA/ LOCATION

LUNGING _____ W T C

WARM-UP _____ W T C

PEAK RIDE _____ W T C

COOL DOWN _____ W T C

MOOD   INSTRUCTOR _____

BEFORE RIDE   DURING RIDE

TAKEAWAYS

_____

_____

IDEAS FOR NEXT RIDE

DAY	TIME	ARENA/ LOCATION

LUNGING _____ W   T   C _____

WARM-UP _____ W   T   C _____

PEAK RIDE _____ W   T   C _____

COOL DOWN _____ W   T   C _____

MOOD   INSTRUCTOR _____

BEFORE RIDE   DURING RIDE

TAKEAWAYS

_____

_____

IDEAS FOR NEXT RIDE

DAY

TIME

ARENA/ LOCATION

LUNGING _____ W   T   C

WARM-UP _____ W   T   C

PEAK RIDE _____ W   T   C

COOL DOWN _____ W   T   C

MOOD

INSTRUCTOR _____

BEFORE RIDE   **DURING RIDE**

TAKEAWAYS

_____

_____

IDEAS FOR NEXT RIDE

DAY	TIME	ARENA/ LOCATION

LUNGING _____ W   T   C

WARM-UP _____ W   T   C

PEAK RIDE _____ W   T   C

COOL DOWN _____ W   T   C

MOOD   **INSTRUCTOR** _____

BEFORE RIDE   **DURING RIDE**

TAKEAWAYS

_____

_____

IDEAS FOR NEXT RIDE

DAY	TIME	ARENA/ LOCATION

LUNGING _____ W   T   C

WARM-UP _____ W   T   C

PEAK RIDE _____ W   T   C

COOL DOWN _____ W   T   C

MOOD

INSTRUCTOR _____

BEFORE RIDE   **DURING RIDE**

TAKEAWAYS

_____

_____

IDEAS FOR NEXT RIDE

DAY

TIME

ARENA/ LOCATION

LUNGING _____ W   T   C _____

WARM-UP _____ W   T   C _____

PEAK RIDE _____ W   T   C _____

COOL DOWN _____ W   T   C _____

MOOD   INSTRUCTOR _____

BEFORE RIDE   DURING RIDE

TAKEAWAYS

_____

_____

IDEAS FOR NEXT RIDE

DAY	TIME	ARENA/ LOCATION

LUNGING_____ W   T   C _____

WARM-UP_____ W   T   C _____

PEAK RIDE_____ W   T   C _____

COOL DOWN_____ W   T   C _____

MOOD   INSTRUCTOR _____

BEFORE RIDE   **DURING RIDE**

TAKEAWAYS

_____

_____

IDEAS FOR NEXT RIDE

DAY

TIME

ARENA/ LOCATION

LUNGING _____ W   T   C

WARM-UP _____ W   T   C

PEAK RIDE _____ W   T   C

COOL DOWN _____ W   T   C

MOOD

INSTRUCTOR _____

BEFORE RIDE   **DURING RIDE**

TAKEAWAYS

_____

_____

IDEAS FOR NEXT RIDE

DAY

TIME

ARENA/ LOCATION

LUNGING _____ W   T   C

WARM-UP _____ W   T   C

PEAK RIDE _____ W   T   C

COOL DOWN _____ W   T   C

MOOD

INSTRUCTOR _____

BEFORE RIDE   DURING RIDE

TAKEAWAYS

_____

_____

IDEAS FOR NEXT RIDE

DAY	TIME	ARENA/ LOCATION

LUNGING _____ W   T   C

WARM-UP _____ W   T   C

PEAK RIDE _____ W   T   C

COOL DOWN _____ W   T   C

MOOD

INSTRUCTOR _____

BEFORE RIDE **DURING RIDE**

TAKEAWAYS

_____

_____

IDEAS FOR NEXT RIDE

DAY

TIME

ARENA/ LOCATION

LUNGING_____ W   T   C

WARM-UP _____ W   T   C

PEAK RIDE _____ W   T   C

COOL DOWN _____ W   T   C

MOOD

INSTRUCTOR _____

BEFORE RIDE   DURING RIDE

TAKEAWAYS

_____

_____

IDEAS FOR NEXT RIDE

DAY

TIME

ARENA/ LOCATION

LUNGING _____ W T C

WARM-UP _____ W T C

PEAK RIDE _____ W T C

COOL DOWN _____ W T C

MOOD

INSTRUCTOR _____

BEFORE RIDE   DURING RIDE

TAKEAWAYS

_____

_____

IDEAS FOR NEXT RIDE

DAY	TIME	ARENA/ LOCATION

LUNGING _____ W   T   C _____

WARM-UP _____ W   T   C _____

PEAK RIDE _____ W   T   C _____

COOL DOWN _____ W   T   C _____

MOOD   **INSTRUCTOR** _____

BEFORE RIDE   DURING RIDE

TAKEAWAYS

_____

_____

IDEAS FOR NEXT RIDE

DAY

TIME

ARENA/ LOCATION

LUNGING _____ W T C

WARM-UP _____ W T C

PEAK RIDE _____ W T C

COOL DOWN _____ W T C

MOOD

INSTRUCTOR _____

BEFORE RIDE   DURING RIDE

TAKEAWAYS

_____

_____

IDEAS FOR NEXT RIDE

DAY

TIME

ARENA/ LOCATION

LUNGING _____ W   T   C

WARM-UP _____ W   T   C

PEAK RIDE _____ W   T   C

COOL DOWN _____ W   T   C

MOOD

INSTRUCTOR _____

BEFORE RIDE   **DURING RIDE**

TAKEAWAYS

_____

_____

IDEAS FOR NEXT RIDE

DAY	TIME	ARENA/ LOCATION

LUNGING _____ W   T   C

WARM-UP _____ W   T   C

PEAK RIDE _____ W   T   C

COOL DOWN _____ W   T   C

MOOD

INSTRUCTOR _____

BEFORE RIDE   **DURING RIDE**

TAKEAWAYS

_____

_____

IDEAS FOR NEXT RIDE

DAY	TIME	ARENA/ LOCATION

LUNGING _____ **W**   **T**   **C** _____

WARM-UP _____ **W**   **T**   **C** _____

PEAK RIDE _____ **W**   **T**   **C** _____

COOL DOWN _____ **W**   **T**   **C** _____

MOOD

INSTRUCTOR _____

BEFORE RIDE   **DURING RIDE**

TAKEAWAYS

_____

_____

IDEAS FOR NEXT RIDE

DAY

TIME

ARENA/ LOCATION

LUNGING _____ W   T   C _____

WARM-UP _____ W   T   C _____

PEAK RIDE _____ W   T   C _____

COOL DOWN _____ W   T   C _____

MOOD   INSTRUCTOR _____

BEFORE RIDE   DURING RIDE

TAKEAWAYS

_____

_____

IDEAS FOR NEXT RIDE

DAY	TIME	ARENA/ LOCATION

LUNGING _____ **W** **T** **C**

WARM-UP _____ **W** **T** **C**

PEAK RIDE _____ **W** **T** **C**

COOL DOWN _____ **W** **T** **C**

MOOD

INSTRUCTOR _____

BEFORE RIDE   DURING RIDE

TAKEAWAYS

_____

_____

IDEAS FOR NEXT RIDE

DAY

TIME

ARENA/ LOCATION

LUNGING _____ W   T   C

WARM-UP _____ W   T   C

PEAK RIDE _____ W   T   C

COOL DOWN _____ W   T   C

MOOD

INSTRUCTOR _____

BEFORE RIDE   **DURING RIDE**

TAKEAWAYS

_____

_____

IDEAS FOR NEXT RIDE

DAY	TIME	ARENA/ LOCATION

LUNGING _____ W   T   C

WARM-UP _____ W   T   C

PEAK RIDE _____ W   T   C

COOL DOWN _____ W   T   C

MOOD   INSTRUCTOR _____

BEFORE RIDE   DURING RIDE

TAKEAWAYS

_____

_____

IDEAS FOR NEXT RIDE

DAY

TIME

ARENA/ LOCATION

LUNGING _____ W T C _____

WARM-UP _____ W T C _____

PEAK RIDE _____ W T C _____

COOL DOWN _____ W T C _____

MOOD

INSTRUCTOR _____

BEFORE RIDE   DURING RIDE

TAKEAWAYS

_____

_____

IDEAS FOR NEXT RIDE

DAY

TIME

ARENA/ LOCATION

LUNGING_____ W T C

WARM-UP_____ W T C

PEAK RIDE_____ W T C

COOL DOWN_____ W T C

MOOD   INSTRUCTOR _____

BEFORE RIDE   **DURING RIDE**

TAKEAWAYS

_____

_____

IDEAS FOR NEXT RIDE

DAY	TIME	ARENA/ LOCATION

LUNGING _____ W   T   C _____

WARM-UP _____ W   T   C _____

PEAK RIDE _____ W   T   C _____

COOL DOWN _____ W   T   C _____

MOOD   **INSTRUCTOR** _____

BEFORE RIDE   DURING RIDE

TAKEAWAYS

_____

_____

IDEAS FOR NEXT RIDE

DAY	TIME	ARENA/ LOCATION

LUNGING _____ W   T   C _____

WARM-UP _____ W   T   C _____

PEAK RIDE _____ W   T   C _____

COOL DOWN _____ W   T   C _____

MOOD

INSTRUCTOR _____

BEFORE RIDE   DURING RIDE

TAKEAWAYS

_____

_____

IDEAS FOR NEXT RIDE

DAY

TIME

ARENA/ LOCATION

LUNGING _____ W T C

WARM-UP _____ W T C

PEAK RIDE _____ W T C

COOL DOWN _____ W T C

MOOD

INSTRUCTOR _____

BEFORE RIDE   **DURING RIDE**

TAKEAWAYS

_____

_____

_____

IDEAS FOR NEXT RIDE

DAY

TIME

ARENA/ LOCATION

LUNGING _____ W   T   C _____

WARM-UP _____ W   T   C _____

PEAK RIDE _____ W   T   C _____

COOL DOWN _____ W   T   C _____

MOOD

INSTRUCTOR _____

BEFORE RIDE   **DURING RIDE**

TAKEAWAYS

_____

_____

IDEAS FOR NEXT RIDE

DAY

TIME

ARENA/ LOCATION

LUNGING _____ W   T   C

WARM-UP _____ W   T   C

PEAK RIDE _____ W   T   C

COOL DOWN _____ W   T   C

MOOD

INSTRUCTOR _____

BEFORE RIDE   **DURING RIDE**

TAKEAWAYS

_____

_____

IDEAS FOR NEXT RIDE

DAY

TIME

ARENA/ LOCATION

LUNGING _____ W   T   C

WARM-UP _____ W   T   C

PEAK RIDE _____ W   T   C

COOL DOWN _____ W   T   C

MOOD

INSTRUCTOR _____

BEFORE RIDE   **DURING RIDE**

TAKEAWAYS

_____

_____

IDEAS FOR NEXT RIDE

DAY

TIME

ARENA/ LOCATION

LUNGING _____ W   T   C _____

WARM-UP _____ W   T   C _____

PEAK RIDE _____ W   T   C _____

COOL DOWN _____ W   T   C _____

MOOD

INSTRUCTOR _____

BEFORE RIDE   **DURING RIDE**

TAKEAWAYS

_____

_____

IDEAS FOR NEXT RIDE

DAY	TIME	ARENA/ LOCATION

LUNGING _____ **W**   **T**   **C** _____

WARM-UP _____ **W**   **T**   **C** _____

PEAK RIDE _____ **W**   **T**   **C** _____

COOL DOWN _____ **W**   **T**   **C** _____

MOOD

INSTRUCTOR _____

BEFORE RIDE   **DURING RIDE**

TAKEAWAYS

_____

_____

IDEAS FOR NEXT RIDE

DAY

TIME

ARENA/ LOCATION

LUNGING _____ W T C

WARM-UP _____ W T C

PEAK RIDE _____ W T C

COOL DOWN _____ W T C

MOOD

INSTRUCTOR _____

BEFORE RIDE **DURING RIDE**

TAKEAWAYS

_____

_____

IDEAS FOR NEXT RIDE

DAY

TIME

ARENA/ LOCATION

LUNGING _____ W   T   C

WARM-UP _____ W   T   C

PEAK RIDE _____ W   T   C

COOL DOWN _____ W   T   C

MOOD

INSTRUCTOR _____

BEFORE RIDE   DURING RIDE

TAKEAWAYS

_____

_____

IDEAS FOR NEXT RIDE

DAY

TIME

ARENA/ LOCATION

LUNGING _____ W   T   C

WARM-UP _____ W   T   C

PEAK RIDE _____ W   T   C

COOL DOWN _____ W   T   C

MOOD

INSTRUCTOR _____

BEFORE RIDE   **DURING RIDE**

TAKEAWAYS

_____

_____

IDEAS FOR NEXT RIDE

DAY

TIME

ARENA/ LOCATION

LUNGING _____ W T C _____

WARM-UP _____ W T C _____

PEAK RIDE _____ W T C _____

COOL DOWN _____ W T C _____

MOOD

INSTRUCTOR _____

BEFORE RIDE   **DURING RIDE**

TAKEAWAYS

_____

_____

IDEAS FOR NEXT RIDE

DAY

TIME

ARENA/ LOCATION

LUNGING _____ W   T   C _____

WARM-UP _____ W   T   C _____

PEAK RIDE _____ W   T   C _____

COOL DOWN _____ W   T   C _____

MOOD

INSTRUCTOR _____

BEFORE RIDE   **DURING RIDE**

TAKEAWAYS

_____

_____

IDEAS FOR NEXT RIDE

DAY	TIME	ARENA/ LOCATION

LUNGING _____ W T C

WARM-UP _____ W T C

PEAK RIDE _____ W T C

COOL DOWN _____ W T C

MOOD   **INSTRUCTOR** _____

BEFORE RIDE   DURING RIDE

TAKEAWAYS

_____

_____

IDEAS FOR NEXT RIDE

DAY

TIME

ARENA/ LOCATION

LUNGING _____ W   T   C _____

WARM-UP _____ W   T   C _____

PEAK RIDE _____ W   T   C _____

COOL DOWN _____ W   T   C _____

MOOD

INSTRUCTOR _____

BEFORE RIDE   **DURING RIDE**

TAKEAWAYS

_____

_____

IDEAS FOR NEXT RIDE

DAY

TIME

ARENA/ LOCATION

LUNGING _____ W   T   C

WARM-UP _____ W   T   C

PEAK RIDE _____ W   T   C

COOL DOWN _____ W   T   C

MOOD

INSTRUCTOR _____

BEFORE RIDE   **DURING RIDE**

TAKEAWAYS

_____

_____

IDEAS FOR NEXT RIDE

DAY

TIME

ARENA/ LOCATION

LUNGING _____ W T C

WARM-UP _____ W T C

PEAK RIDE _____ W T C

COOL DOWN _____ W T C

MOOD

INSTRUCTOR _____

BEFORE RIDE   DURING RIDE

TAKEAWAYS

_____

_____

IDEAS FOR NEXT RIDE

DAY	TIME	ARENA/ LOCATION

LUNGING_____ W   T   C _____

WARM-UP _____ W   T   C _____

PEAK RIDE _____ W   T   C _____

COOL DOWN _____ W   T   C _____

MOOD   INSTRUCTOR _____

BEFORE RIDE   DURING RIDE

TAKEAWAYS

_____

_____

IDEAS FOR NEXT RIDE

DAY

TIME

ARENA/ LOCATION

LUNGING _____ W   T   C

WARM-UP _____ W   T   C

PEAK RIDE _____ W   T   C

COOL DOWN _____ W   T   C

MOOD

INSTRUCTOR _____

BEFORE RIDE   **DURING RIDE**

TAKEAWAYS

_____

_____

IDEAS FOR NEXT RIDE

DAY	TIME	ARENA/ LOCATION

LUNGING _____ W   T   C

WARM-UP _____ W   T   C

PEAK RIDE _____ W   T   C

COOL DOWN _____ W   T   C

MOOD

INSTRUCTOR _____

BEFORE RIDE   DURING RIDE

TAKEAWAYS

_____

_____

IDEAS FOR NEXT RIDE

DAY

TIME

ARENA/ LOCATION

LUNGING _____ W   T   C _____

WARM-UP _____ W   T   C _____

PEAK RIDE _____ W   T   C _____

COOL DOWN _____ W   T   C _____

MOOD

INSTRUCTOR _____

BEFORE RIDE   DURING RIDE

TAKEAWAYS

_____

_____

IDEAS FOR NEXT RIDE

DAY

TIME

ARENA/ LOCATION

LUNGING _____ W   T   C

WARM-UP _____ W   T   C

PEAK RIDE _____ W   T   C

COOL DOWN _____ W   T   C

MOOD   INSTRUCTOR _____

BEFORE RIDE   DURING RIDE

TAKEAWAYS

_____

_____

IDEAS FOR NEXT RIDE

DAY

TIME

ARENA/ LOCATION

LUNGING _____ W   T   C

WARM-UP _____ W   T   C

PEAK RIDE _____ W   T   C

COOL DOWN _____ W   T   C

MOOD

INSTRUCTOR _____

BEFORE RIDE   DURING RIDE

TAKEAWAYS

_____

_____

IDEAS FOR NEXT RIDE

DAY	TIME	ARENA/ LOCATION

LUNGING_____ W   T   C_____

WARM-UP_____ W   T   C_____

PEAK RIDE_____ W   T   C_____

COOL DOWN_____ W   T   C_____

MOOD

INSTRUCTOR _____

BEFORE RIDE   **DURING RIDE**

TAKEAWAYS

_____

_____

IDEAS FOR NEXT RIDE

DAY

TIME

ARENA/ LOCATION

LUNGING _____ W   T   C

WARM-UP _____ W   T   C

PEAK RIDE _____ W   T   C

COOL DOWN _____ W   T   C

MOOD

INSTRUCTOR _____

BEFORE RIDE   **DURING RIDE**

TAKEAWAYS

_____

_____

IDEAS FOR NEXT RIDE

DAY

TIME

ARENA/ LOCATION

LUNGING _____ W　　T　　C

WARM-UP _____ W　　T　　C

PEAK RIDE _____ W　　T　　C

COOL DOWN _____ W　　T　　C

MOOD

INSTRUCTOR _____

BEFORE RIDE　　DURING RIDE

TAKEAWAYS

_____

_____

IDEAS FOR NEXT RIDE

DAY	TIME	ARENA/ LOCATION

LUNGING _____ W   T   C

WARM-UP _____ W   T   C

PEAK RIDE _____ W   T   C

COOL DOWN _____ W   T   C

MOOD

INSTRUCTOR _____

BEFORE RIDE   DURING RIDE

TAKEAWAYS

_____

_____

IDEAS FOR NEXT RIDE

DAY

TIME

ARENA/ LOCATION

LUNGING _____ W   T   C

WARM-UP _____ W   T   C

PEAK RIDE _____ W   T   C

COOL DOWN _____ W   T   C

MOOD

INSTRUCTOR _____

BEFORE RIDE   DURING RIDE

TAKEAWAYS

_____

_____

IDEAS FOR NEXT RIDE

DAY

TIME

ARENA/ LOCATION

LUNGING _____ W   T   C

WARM-UP _____ W   T   C

PEAK RIDE _____ W   T   C

COOL DOWN _____ W   T   C

MOOD

INSTRUCTOR _____

BEFORE RIDE   **DURING RIDE**

TAKEAWAYS

_____

_____

IDEAS FOR NEXT RIDE

DAY	TIME	ARENA/ LOCATION

LUNGING _____ **W**   **T**   **C** _____

WARM-UP _____ **W**   **T**   **C** _____

PEAK RIDE _____ **W**   **T**   **C** _____

COOL DOWN _____ **W**   **T**   **C** _____

MOOD

INSTRUCTOR _____

BEFORE RIDE   **DURING RIDE**

TAKEAWAYS

_____

_____

IDEAS FOR NEXT RIDE

DAY

TIME

ARENA/ LOCATION

LUNGING _____ W   T   C

WARM-UP _____ W   T   C

PEAK RIDE _____ W   T   C

COOL DOWN _____ W   T   C

MOOD

INSTRUCTOR _____

BEFORE RIDE   **DURING RIDE**

TAKEAWAYS

_____

_____

IDEAS FOR NEXT RIDE

DAY	TIME	ARENA/ LOCATION

LUNGING _____ W   T   C

WARM-UP _____ W   T   C

PEAK RIDE _____ W   T   C

COOL DOWN _____ W   T   C

MOOD

INSTRUCTOR _____

BEFORE RIDE   **DURING RIDE**

TAKEAWAYS

_____

_____

IDEAS FOR NEXT RIDE

DAY

TIME

ARENA/ LOCATION

LUNGING _____ W   T   C

WARM-UP _____ W   T   C

PEAK RIDE _____ W   T   C

COOL DOWN _____ W   T   C

MOOD

INSTRUCTOR _____

BEFORE RIDE   DURING RIDE

TAKEAWAYS

_____

_____

IDEAS FOR NEXT RIDE

DAY

TIME

ARENA/ LOCATION

LUNGING _____ W   T   C

WARM-UP _____ W   T   C

PEAK RIDE _____ W   T   C

COOL DOWN _____ W   T   C

MOOD   INSTRUCTOR _____

BEFORE RIDE   DURING RIDE

TAKEAWAYS

_____

_____

IDEAS FOR NEXT RIDE

DAY

TIME

ARENA/ LOCATION

LUNGING _____ W T C

WARM-UP _____ W T C

PEAK RIDE _____ W T C

COOL DOWN _____ W T C

MOOD

INSTRUCTOR _____

BEFORE RIDE   **DURING RIDE**

TAKEAWAYS

_____

_____

IDEAS FOR NEXT RIDE

DAY

TIME

ARENA/ LOCATION

LUNGING _____ W   T   C _____

WARM-UP _____ W   T   C _____

PEAK RIDE _____ W   T   C _____

COOL DOWN _____ W   T   C _____

MOOD

INSTRUCTOR _____

BEFORE RIDE   **DURING RIDE**

TAKEAWAYS

_____

_____

IDEAS FOR NEXT RIDE

DAY

TIME

ARENA/ LOCATION

LUNGING _____ W   T   C

WARM-UP _____ W   T   C

PEAK RIDE _____ W   T   C

COOL DOWN _____ W   T   C

MOOD

INSTRUCTOR _____

BEFORE RIDE   **DURING RIDE**

TAKEAWAYS

_____

_____

IDEAS FOR NEXT RIDE

DAY	TIME	ARENA/ LOCATION

LUNGING_____ W   T   C _____

WARM-UP _____ W   T   C _____

PEAK RIDE _____ W   T   C _____

COOL DOWN _____ W   T   C _____

MOOD

INSTRUCTOR _____

BEFORE RIDE   **DURING RIDE**

TAKEAWAYS

_____

_____

_____

IDEAS FOR NEXT RIDE

DAY

TIME

ARENA/ LOCATION

LUNGING _____ W   T   C

WARM-UP _____ W   T   C

PEAK RIDE _____ W   T   C

COOL DOWN _____ W   T   C

MOOD

INSTRUCTOR _____

BEFORE RIDE   **DURING RIDE**

TAKEAWAYS

_____

_____

IDEAS FOR NEXT RIDE

DAY	TIME	ARENA/ LOCATION

LUNGING _____ W   T   C

WARM-UP _____ W   T   C

PEAK RIDE _____ W   T   C

COOL DOWN _____ W   T   C

MOOD   INSTRUCTOR _____

BEFORE RIDE   DURING RIDE

TAKEAWAYS

_____

_____

IDEAS FOR NEXT RIDE

DAY

TIME

ARENA/ LOCATION

LUNGING _____ W   T   C

WARM-UP _____ W   T   C

PEAK RIDE _____ W   T   C

COOL DOWN _____ W   T   C

MOOD

INSTRUCTOR _____

BEFORE RIDE   **DURING RIDE**

TAKEAWAYS

_____

_____

IDEAS FOR NEXT RIDE

DAY

TIME

ARENA/ LOCATION

LUNGING _____ W T C _____

WARM-UP _____ W T C _____

PEAK RIDE _____ W T C _____

COOL DOWN _____ W T C _____

MOOD

INSTRUCTOR _____

BEFORE RIDE   **DURING RIDE**

TAKEAWAYS

_____

_____

IDEAS FOR NEXT RIDE

DAY

TIME

ARENA/ LOCATION

LUNGING _____ W T C

WARM-UP _____ W T C

PEAK RIDE _____ W T C

COOL DOWN _____ W T C

MOOD

INSTRUCTOR _____

BEFORE RIDE   DURING RIDE

TAKEAWAYS

_____

_____

IDEAS FOR NEXT RIDE

DAY	TIME	ARENA/ LOCATION

LUNGING _____ W   T   C

WARM-UP _____ W   T   C

PEAK RIDE _____ W   T   C

COOL DOWN _____ W   T   C

MOOD   **INSTRUCTOR** _____

BEFORE RIDE   DURING RIDE

TAKEAWAYS

_____

_____

IDEAS FOR NEXT RIDE

DAY

TIME

ARENA/ LOCATION

LUNGING _____ W   T   C

WARM-UP _____ W   T   C

PEAK RIDE _____ W   T   C

COOL DOWN _____ W   T   C

MOOD   **INSTRUCTOR** _____

BEFORE RIDE   **DURING RIDE**

TAKEAWAYS

_____

_____

IDEAS FOR NEXT RIDE

DAY

TIME

ARENA/ LOCATION

LUNGING_____ W   T   C _____

WARM-UP _____ W   T   C _____

PEAK RIDE _____ W   T   C _____

COOL DOWN _____ W   T   C _____

MOOD

INSTRUCTOR _____

BEFORE RIDE   **DURING RIDE**

TAKEAWAYS

_____

_____

IDEAS FOR NEXT RIDE

DAY

TIME

ARENA/ LOCATION

LUNGING _____ W   T   C

WARM-UP _____ W   T   C

PEAK RIDE _____ W   T   C

COOL DOWN _____ W   T   C

MOOD   INSTRUCTOR _____

BEFORE RIDE   **DURING RIDE**

TAKEAWAYS

_____

_____

IDEAS FOR NEXT RIDE

DAY

TIME

ARENA/ LOCATION

LUNGING _____ W   T   C

WARM-UP _____ W   T   C

PEAK RIDE _____ W   T   C

COOL DOWN _____ W   T   C

MOOD

INSTRUCTOR _____

BEFORE RIDE   DURING RIDE

TAKEAWAYS

_____

_____

IDEAS FOR NEXT RIDE

DAY

TIME

ARENA/ LOCATION

LUNGING _____ W   T   C

WARM-UP _____ W   T   C

PEAK RIDE _____ W   T   C

COOL DOWN _____ W   T   C

MOOD

INSTRUCTOR _____

BEFORE RIDE   DURING RIDE

TAKEAWAYS

_____

_____

IDEAS FOR NEXT RIDE

DAY	TIME	ARENA/ LOCATION

LUNGING _____ W   T   C

WARM-UP _____ W   T   C

PEAK RIDE _____ W   T   C

COOL DOWN _____ W   T   C

MOOD

INSTRUCTOR _____

BEFORE RIDE   DURING RIDE

TAKEAWAYS

_____

_____

IDEAS FOR NEXT RIDE

DAY

TIME

ARENA/ LOCATION

LUNGING _____ **W**   **T**   **C**

WARM-UP _____ **W**   **T**   **C**

PEAK RIDE _____ **W**   **T**   **C**

COOL DOWN _____ **W**   **T**   **C**

MOOD

INSTRUCTOR _____

BEFORE RIDE   **DURING RIDE**

TAKEAWAYS

_____

_____

IDEAS FOR NEXT RIDE

DAY

TIME

ARENA/ LOCATION

LUNGING _____ W   T   C _____

WARM-UP _____ W   T   C _____

PEAK RIDE _____ W   T   C _____

COOL DOWN _____ W   T   C _____

MOOD   **INSTRUCTOR** _____

BEFORE RIDE   DURING RIDE

TAKEAWAYS

_____

_____

IDEAS FOR NEXT RIDE

DAY

TIME

ARENA/ LOCATION

LUNGING _____ W   T   C

WARM-UP _____ W   T   C

PEAK RIDE _____ W   T   C

COOL DOWN _____ W   T   C

MOOD

INSTRUCTOR _____

BEFORE RIDE   **DURING RIDE**

TAKEAWAYS

_____

_____

IDEAS FOR NEXT RIDE

DAY

TIME

ARENA/ LOCATION

LUNGING _____ W   T   C _____

WARM-UP _____ W   T   C _____

PEAK RIDE _____ W   T   C _____

COOL DOWN _____ W   T   C _____

MOOD

INSTRUCTOR _____

BEFORE RIDE   **DURING RIDE**

TAKEAWAYS

_____

_____

IDEAS FOR NEXT RIDE

DAY

TIME

ARENA/ LOCATION

LUNGING _____ W   T   C

WARM-UP _____ W   T   C

PEAK RIDE _____ W   T   C

COOL DOWN _____ W   T   C

MOOD

INSTRUCTOR _____

BEFORE RIDE   DURING RIDE

TAKEAWAYS

_____

_____

IDEAS FOR NEXT RIDE

DAY	**TIME**	**ARENA/ LOCATION**

LUNGING_____ **W**   **T**   **C**

WARM-UP _____ **W**   **T**   **C**

PEAK RIDE_____ **W**   **T**   **C**

COOL DOWN _____ **W**   **T**   **C**

MOOD   **INSTRUCTOR** _____

BEFORE RIDE   **DURING RIDE**

TAKEAWAYS

_____

_____

IDEAS FOR NEXT RIDE

DAY

TIME

ARENA/ LOCATION

LUNGING _____ W   T   C

WARM-UP _____ W   T   C

PEAK RIDE _____ W   T   C

COOL DOWN _____ W   T   C

MOOD

INSTRUCTOR _____

BEFORE RIDE   **DURING RIDE**

TAKEAWAYS

_____

_____

IDEAS FOR NEXT RIDE

DAY

TIME

ARENA/ LOCATION

LUNGING _____ W   T   C _____

WARM-UP _____ W   T   C _____

PEAK RIDE _____ W   T   C _____

COOL DOWN _____ W   T   C _____

MOOD

INSTRUCTOR _____

BEFORE RIDE   **DURING RIDE**

TAKEAWAYS

_____

_____

IDEAS FOR NEXT RIDE

DAY

TIME

ARENA/ LOCATION

LUNGING _____ W   T   C

WARM-UP _____ W   T   C

PEAK RIDE _____ W   T   C

COOL DOWN _____ W   T   C

MOOD

INSTRUCTOR _____

BEFORE RIDE   DURING RIDE

TAKEAWAYS

_____

_____

IDEAS FOR NEXT RIDE

DAY	TIME	ARENA/ LOCATION

LUNGING _____ W   T   C _____

WARM-UP _____ W   T   C _____

PEAK RIDE _____ W   T   C _____

COOL DOWN _____ W   T   C _____

MOOD   **INSTRUCTOR** _____

BEFORE RIDE   **DURING RIDE**

TAKEAWAYS

_____

_____

IDEAS FOR NEXT RIDE

DAY

TIME

ARENA/ LOCATION

LUNGING _____ **W** **T** **C** _____

WARM-UP _____ **W** **T** **C** _____

PEAK RIDE _____ **W** **T** **C** _____

COOL DOWN _____ **W** **T** **C** _____

MOOD

INSTRUCTOR _____

BEFORE RIDE   **DURING RIDE**

TAKEAWAYS

_____

_____

IDEAS FOR NEXT RIDE

DAY	TIME	ARENA/ LOCATION

LUNGING _____ **W**   **T**   **C** _____

WARM-UP _____ **W**   **T**   **C** _____

PEAK RIDE _____ **W**   **T**   **C** _____

COOL DOWN _____ **W**   **T**   **C** _____

MOOD

INSTRUCTOR _____

BEFORE RIDE   **DURING RIDE**

TAKEAWAYS

_____

_____

_____

IDEAS FOR NEXT RIDE

DAY	TIME	ARENA/ LOCATION

LUNGING _____ W T C

WARM-UP _____ W T C

PEAK RIDE _____ W T C

COOL DOWN _____ W T C

MOOD

INSTRUCTOR _____

BEFORE RIDE   DURING RIDE

TAKEAWAYS

_____

_____

IDEAS FOR NEXT RIDE

DAY

TIME

ARENA/ LOCATION

LUNGING _____ W   T   C

WARM-UP _____ W   T   C

PEAK RIDE _____ W   T   C

COOL DOWN _____ W   T   C

MOOD

INSTRUCTOR _____

BEFORE RIDE   DURING RIDE

TAKEAWAYS

_____

_____

IDEAS FOR NEXT RIDE

DAY

TIME

ARENA/ LOCATION

LUNGING _____ W   T   C _____

WARM-UP _____ W   T   C _____

PEAK RIDE _____ W   T   C _____

COOL DOWN _____ W   T   C _____

MOOD

INSTRUCTOR _____

BEFORE RIDE   DURING RIDE

TAKEAWAYS

_____

_____

IDEAS FOR NEXT RIDE

DAY	TIME	ARENA/ LOCATION

LUNGING _____ **W** **T** **C** _____

WARM-UP _____ **W** **T** **C** _____

PEAK RIDE _____ **W** **T** **C** _____

COOL DOWN _____ **W** **T** **C** _____

MOOD   **INSTRUCTOR** _____

BEFORE RIDE   **DURING RIDE**

TAKEAWAYS

_____

_____

IDEAS FOR NEXT RIDE

DAY

TIME

ARENA/ LOCATION

LUNGING _____ W T C

WARM-UP _____ W T C

PEAK RIDE _____ W T C

COOL DOWN _____ W T C

MOOD

INSTRUCTOR _____

BEFORE RIDE   **DURING RIDE**

TAKEAWAYS

_____

_____

IDEAS FOR NEXT RIDE

DAY	TIME	ARENA/ LOCATION

LUNGING _____ W   T   C

WARM-UP _____ W   T   C

PEAK RIDE _____ W   T   C

COOL DOWN _____ W   T   C

MOOD   **INSTRUCTOR** _____

BEFORE RIDE   **DURING RIDE**

TAKEAWAYS

_____

_____

IDEAS FOR NEXT RIDE

DAY

TIME

ARENA/ LOCATION

LUNGING _____ W   T   C _____

WARM-UP _____ W   T   C _____

PEAK RIDE _____ W   T   C _____

COOL DOWN _____ W   T   C _____

MOOD

INSTRUCTOR _____

BEFORE RIDE   DURING RIDE

TAKEAWAYS

_____

_____

IDEAS FOR NEXT RIDE

DAY

TIME

ARENA/ LOCATION

LUNGING_____ W   T   C

WARM-UP _____ W   T   C

PEAK RIDE _____ W   T   C

COOL DOWN _____ W   T   C

MOOD   INSTRUCTOR _____

BEFORE RIDE   DURING RIDE

TAKEAWAYS

_____

_____

IDEAS FOR NEXT RIDE

DAY

TIME

ARENA/ LOCATION

LUNGING _____ W　T　C

WARM-UP _____ W　T　C

PEAK RIDE _____ W　T　C

COOL DOWN _____ W　T　C

MOOD

INSTRUCTOR _____

BEFORE RIDE　　**DURING RIDE**

TAKEAWAYS

_____

_____

IDEAS FOR NEXT RIDE

DAY

TIME

ARENA/ LOCATION

LUNGING _____ W   T   C _____

WARM-UP _____ W   T   C _____

PEAK RIDE _____ W   T   C _____

COOL DOWN _____ W   T   C _____

MOOD   **INSTRUCTOR** _____

BEFORE RIDE   **DURING RIDE**

TAKEAWAYS

_____

_____

IDEAS FOR NEXT RIDE

DAY

TIME

ARENA/ LOCATION

LUNGING _____ W T C

WARM-UP _____ W T C

PEAK RIDE _____ W T C

COOL DOWN _____ W T C

MOOD

INSTRUCTOR _____

BEFORE RIDE   DURING RIDE

TAKEAWAYS

_____

_____

IDEAS FOR NEXT RIDE

DAY	TIME	ARENA/ LOCATION

LUNGING_____ W   T   C _____

WARM-UP _____ W   T   C _____

PEAK RIDE _____ W   T   C _____

COOL DOWN _____ W   T   C _____

MOOD

INSTRUCTOR _____

BEFORE RIDE   **DURING RIDE**

TAKEAWAYS

_____

_____

IDEAS FOR NEXT RIDE

DAY	TIME	ARENA/ LOCATION

LUNGING _____ W   T   C _____

WARM-UP _____ W   T   C _____

PEAK RIDE _____ W   T   C _____

COOL DOWN _____ W   T   C _____

MOOD

INSTRUCTOR _____

BEFORE RIDE   **DURING RIDE**

TAKEAWAYS

_____

_____

IDEAS FOR NEXT RIDE

DAY	TIME	ARENA/ LOCATION

LUNGING _____ W   T   C _____

WARM-UP _____ W   T   C _____

PEAK RIDE _____ W   T   C _____

COOL DOWN _____ W   T   C _____

MOOD

INSTRUCTOR _____

BEFORE RIDE   **DURING RIDE**

TAKEAWAYS

_____

_____

IDEAS FOR NEXT RIDE

DAY

TIME

ARENA/ LOCATION

LUNGING _____ W T C

WARM-UP _____ W T C

PEAK RIDE _____ W T C

COOL DOWN _____ W T C

MOOD

INSTRUCTOR _____

BEFORE RIDE **DURING RIDE**

TAKEAWAYS

_____

_____

IDEAS FOR NEXT RIDE

DAY

TIME

ARENA/ LOCATION

LUNGING _____ W   T   C _____

WARM-UP _____ W   T   C _____

PEAK RIDE _____ W   T   C _____

COOL DOWN _____ W   T   C _____

MOOD   INSTRUCTOR _____

BEFORE RIDE   **DURING RIDE**

TAKEAWAYS

_____

_____

IDEAS FOR NEXT RIDE

DAY

TIME

ARENA/ LOCATION

LUNGING _____ **W** **T** **C**

WARM-UP _____ **W** **T** **C**

PEAK RIDE _____ **W** **T** **C**

COOL DOWN _____ **W** **T** **C**

MOOD

INSTRUCTOR _____

BEFORE RIDE   **DURING RIDE**

TAKEAWAYS

_____

_____

IDEAS FOR NEXT RIDE

DAY	TIME	ARENA/ LOCATION

LUNGING _____ W   T   C

WARM-UP _____ W   T   C

PEAK RIDE _____ W   T   C

COOL DOWN _____ W   T   C

MOOD   **INSTRUCTOR** _____

BEFORE RIDE   DURING RIDE

TAKEAWAYS

_____

_____

_____

IDEAS FOR NEXT RIDE

DAY	TIME	ARENA/ LOCATION

LUNGING _____ W   T   C _____

WARM-UP _____ W   T   C _____

PEAK RIDE _____ W   T   C _____

COOL DOWN _____ W   T   C _____

MOOD   INSTRUCTOR _____

BEFORE RIDE   DURING RIDE

TAKEAWAYS

_____

_____

IDEAS FOR NEXT RIDE

DAY

TIME

ARENA/ LOCATION

LUNGING _____ W   T   C

WARM-UP _____ W   T   C

PEAK RIDE _____ W   T   C

COOL DOWN _____ W   T   C

MOOD

INSTRUCTOR _____

BEFORE RIDE   DURING RIDE

TAKEAWAYS

_____

_____

IDEAS FOR NEXT RIDE

DAY

TIME

ARENA/ LOCATION

LUNGING _____ W T C

WARM-UP _____ W T C

PEAK RIDE _____ W T C

COOL DOWN _____ W T C

MOOD

INSTRUCTOR _____

BEFORE RIDE   DURING RIDE

TAKEAWAYS

_____

_____

IDEAS FOR NEXT RIDE

DAY

TIME

ARENA/ LOCATION

LUNGING _____ W   T   C

WARM-UP _____ W   T   C

PEAK RIDE _____ W   T   C

COOL DOWN _____ W   T   C

MOOD

INSTRUCTOR _____

BEFORE RIDE   DURING RIDE

TAKEAWAYS

_____

_____

IDEAS FOR NEXT RIDE

DAY	TIME	ARENA/ LOCATION

LUNGING _____ W   T   C

WARM-UP _____ W   T   C

PEAK RIDE _____ W   T   C

COOL DOWN _____ W   T   C

MOOD

INSTRUCTOR _____

BEFORE RIDE   DURING RIDE

TAKEAWAYS

_____

_____

IDEAS FOR NEXT RIDE

DAY	TIME	ARENA/ LOCATION

LUNGING _____ W   T   C

WARM-UP _____ W   T   C

PEAK RIDE _____ W   T   C

COOL DOWN _____ W   T   C

MOOD

INSTRUCTOR _____

BEFORE RIDE   **DURING RIDE**

TAKEAWAYS

_____

_____

IDEAS FOR NEXT RIDE

DAY

TIME

ARENA/ LOCATION

LUNGING _____ W   T   C

WARM-UP _____ W   T   C

PEAK RIDE _____ W   T   C

COOL DOWN _____ W   T   C

MOOD   INSTRUCTOR _____

BEFORE RIDE   DURING RIDE

TAKEAWAYS

_____

_____

IDEAS FOR NEXT RIDE

DAY

TIME

ARENA/ LOCATION

LUNGING _____ W   T   C

WARM-UP _____ W   T   C

PEAK RIDE _____ W   T   C

COOL DOWN _____ W   T   C

MOOD

INSTRUCTOR _____

BEFORE RIDE   DURING RIDE

TAKEAWAYS

_____

_____

IDEAS FOR NEXT RIDE

DAY

TIME

ARENA/ LOCATION

LUNGING _____ W   T   C

WARM-UP _____ W   T   C

PEAK RIDE _____ W   T   C

COOL DOWN _____ W   T   C

MOOD

INSTRUCTOR _____

BEFORE RIDE   **DURING RIDE**

TAKEAWAYS

_____

_____

IDEAS FOR NEXT RIDE

DAY

TIME

ARENA/ LOCATION

LUNGING _____ W   T   C

WARM-UP _____ W   T   C

PEAK RIDE _____ W   T   C

COOL DOWN _____ W   T   C

MOOD

INSTRUCTOR _____

BEFORE RIDE   **DURING RIDE**

TAKEAWAYS

_____

_____

IDEAS FOR NEXT RIDE

DAY

TIME

ARENA/ LOCATION

LUNGING _____ W   T   C

WARM-UP _____ W   T   C

PEAK RIDE _____ W   T   C

COOL DOWN _____ W   T   C

MOOD   INSTRUCTOR _____

BEFORE RIDE   DURING RIDE

TAKEAWAYS

_____

_____

IDEAS FOR NEXT RIDE

DAY	TIME	ARENA/ LOCATION

LUNGING _____ W T C

WARM-UP _____ W T C

PEAK RIDE _____ W T C

COOL DOWN _____ W T C

MOOD

INSTRUCTOR _____

BEFORE RIDE   **DURING RIDE**

TAKEAWAYS

_____

_____

IDEAS FOR NEXT RIDE

DAY

TIME

ARENA/ LOCATION

LUNGING _____ W   T   C _____

WARM-UP _____ W   T   C _____

PEAK RIDE _____ W   T   C _____

COOL DOWN _____ W   T   C _____

MOOD

INSTRUCTOR _____

BEFORE RIDE   **DURING RIDE**

TAKEAWAYS

_____

_____

IDEAS FOR NEXT RIDE

DAY

TIME

ARENA/ LOCATION

LUNGING _____ W   T   C

WARM-UP _____ W   T   C

PEAK RIDE _____ W   T   C

COOL DOWN _____ W   T   C

MOOD

INSTRUCTOR _____

BEFORE RIDE   DURING RIDE

TAKEAWAYS

_____

_____

IDEAS FOR NEXT RIDE

DAY

TIME

ARENA/ LOCATION

LUNGING _____ W   T   C

WARM-UP _____ W   T   C

PEAK RIDE _____ W   T   C

COOL DOWN _____ W   T   C

MOOD   INSTRUCTOR _____

BEFORE RIDE   DURING RIDE

TAKEAWAYS

_____

_____

IDEAS FOR NEXT RIDE

DAY

TIME

ARENA/ LOCATION

LUNGING _____ W   T   C

WARM-UP _____ W   T   C

PEAK RIDE _____ W   T   C

COOL DOWN _____ W   T   C

MOOD

INSTRUCTOR _____

BEFORE RIDE   **DURING RIDE**

TAKEAWAYS

_____

_____

IDEAS FOR NEXT RIDE

DAY

TIME

ARENA/ LOCATION

LUNGING _____ W T C _____

WARM-UP _____ W T C _____

PEAK RIDE _____ W T C _____

COOL DOWN _____ W T C _____

MOOD

INSTRUCTOR _____

BEFORE RIDE   DURING RIDE

TAKEAWAYS

_____

_____

IDEAS FOR NEXT RIDE

DAY	TIME	ARENA/ LOCATION

LUNGING _____ W   T   C _____

WARM-UP _____ W   T   C _____

PEAK RIDE _____ W   T   C _____

COOL DOWN _____ W   T   C _____

MOOD

INSTRUCTOR _____

BEFORE RIDE   **DURING RIDE**

TAKEAWAYS

_____

_____

IDEAS FOR NEXT RIDE

DAY	TIME	ARENA/ LOCATION

LUNGING _____ W T C

WARM-UP _____ W T C

PEAK RIDE _____ W T C

COOL DOWN _____ W T C

MOOD

INSTRUCTOR _____

BEFORE RIDE   DURING RIDE

TAKEAWAYS

_____

_____

IDEAS FOR NEXT RIDE

DAY | **TIME** | **ARENA/ LOCATION**

LUNGING _____ W   T   C _____

WARM-UP _____ W   T   C _____

PEAK RIDE _____ W   T   C _____

COOL DOWN _____ W   T   C _____

MOOD

INSTRUCTOR _____

BEFORE RIDE   **DURING RIDE**

TAKEAWAYS

_____

_____

IDEAS FOR NEXT RIDE

DAY

TIME

ARENA/ LOCATION

LUNGING _____ W   T   C

WARM-UP _____ W   T   C

PEAK RIDE _____ W   T   C

COOL DOWN _____ W   T   C

MOOD

INSTRUCTOR _____

BEFORE RIDE   **DURING RIDE**

TAKEAWAYS

_____

_____

IDEAS FOR NEXT RIDE

DAY	TIME	ARENA/ LOCATION

LUNGING _____ **W**   **T**   **C** _____

WARM-UP _____ **W**   **T**   **C** _____

PEAK RIDE _____ **W**   **T**   **C** _____

COOL DOWN _____ **W**   **T**   **C** _____

MOOD

INSTRUCTOR _____

BEFORE RIDE   DURING RIDE

TAKEAWAYS

_____

_____

IDEAS FOR NEXT RIDE

DAY

TIME

ARENA/ LOCATION

LUNGING _____ W   T   C

WARM-UP _____ W   T   C

PEAK RIDE _____ W   T   C

COOL DOWN _____ W   T   C

MOOD

INSTRUCTOR _____

BEFORE RIDE   **DURING RIDE**

TAKEAWAYS

_____

_____

IDEAS FOR NEXT RIDE

DAY	TIME	ARENA/ LOCATION

LUNGING _____ W T C

WARM-UP _____ W T C

PEAK RIDE _____ W T C

COOL DOWN _____ W T C

MOOD

INSTRUCTOR _____

BEFORE RIDE   **DURING RIDE**

TAKEAWAYS

_____

_____

IDEAS FOR NEXT RIDE

DAY	TIME	ARENA/ LOCATION

LUNGING _____ W   T   C

WARM-UP _____ W   T   C

PEAK RIDE _____ W   T   C

COOL DOWN _____ W   T   C

MOOD

INSTRUCTOR _____

BEFORE RIDE **DURING RIDE**

TAKEAWAYS

_____

_____

IDEAS FOR NEXT RIDE

DAY	TIME	ARENA/ LOCATION

LUNGING _____ W   T   C _____

WARM-UP _____ W   T   C _____

PEAK RIDE _____ W   T   C _____

COOL DOWN _____ W   T   C _____

MOOD   INSTRUCTOR _____

BEFORE RIDE   DURING RIDE

TAKEAWAYS

_____

_____

IDEAS FOR NEXT RIDE

DAY	TIME	ARENA/ LOCATION

LUNGING _____ W   T   C

WARM-UP _____ W   T   C

PEAK RIDE _____ W   T   C

COOL DOWN _____ W   T   C

MOOD   **INSTRUCTOR** _____

BEFORE RIDE   **DURING RIDE**

TAKEAWAYS

_____

_____

IDEAS FOR NEXT RIDE

DAY

TIME

ARENA/ LOCATION

LUNGING_____ W T C

WARM-UP _____ W T C

PEAK RIDE _____ W T C

COOL DOWN _____ W T C

MOOD   INSTRUCTOR _____

BEFORE RIDE   **DURING RIDE**

TAKEAWAYS

_____

_____

IDEAS FOR NEXT RIDE

DAY

TIME

ARENA/ LOCATION

LUNGING _____ W T C

WARM-UP _____ W T C

PEAK RIDE _____ W T C

COOL DOWN _____ W T C

MOOD

INSTRUCTOR _____

BEFORE RIDE   **DURING RIDE**

TAKEAWAYS

_____

_____

IDEAS FOR NEXT RIDE

DAY

TIME

ARENA/ LOCATION

LUNGING _____ W   T   C

WARM-UP _____ W   T   C

PEAK RIDE _____ W   T   C

COOL DOWN _____ W   T   C

MOOD

INSTRUCTOR _____

BEFORE RIDE   **DURING RIDE**

TAKEAWAYS

_____

_____

IDEAS FOR NEXT RIDE

DAY	TIME	ARENA/ LOCATION

LUNGING _____ **W** **T** **C**

WARM-UP _____ **W** **T** **C**

PEAK RIDE _____ **W** **T** **C**

COOL DOWN _____ **W** **T** **C**

MOOD

INSTRUCTOR _____

BEFORE RIDE   DURING RIDE

TAKEAWAYS

_____

_____

IDEAS FOR NEXT RIDE

DAY	TIME	ARENA/ LOCATION

LUNGING_____ W   T   C

WARM-UP _____ W   T   C

PEAK RIDE _____ W   T   C

COOL DOWN _____ W   T   C

MOOD   INSTRUCTOR _____

BEFORE RIDE   **DURING RIDE**

TAKEAWAYS

_____

_____

IDEAS FOR NEXT RIDE

DAY	TIME	ARENA/ LOCATION

LUNGING _____ W   T   C

WARM-UP _____ W   T   C

PEAK RIDE _____ W   T   C

COOL DOWN _____ W   T   C

MOOD　　　**INSTRUCTOR** _____

BEFORE RIDE　　**DURING RIDE**

TAKEAWAYS

_____

_____

IDEAS FOR NEXT RIDE

DAY

TIME

ARENA/ LOCATION

LUNGING _____ W T C

WARM-UP _____ W T C

PEAK RIDE _____ W T C

COOL DOWN _____ W T C

MOOD

INSTRUCTOR _____

BEFORE RIDE   DURING RIDE

TAKEAWAYS

_____

_____

IDEAS FOR NEXT RIDE

DAY

TIME

ARENA/ LOCATION

LUNGING_____ W   T   C _____

WARM-UP _____ W   T   C _____

PEAK RIDE _____ W   T   C _____

COOL DOWN _____ W   T   C _____

MOOD   **INSTRUCTOR** _____

BEFORE RIDE   DURING RIDE

TAKEAWAYS

_____

_____

IDEAS FOR NEXT RIDE

DAY	TIME	ARENA/ LOCATION

LUNGING_____ W T C

WARM-UP _____ W T C

PEAK RIDE _____ W T C

COOL DOWN_____ W T C

MOOD

INSTRUCTOR _____

BEFORE RIDE   **DURING RIDE**

TAKEAWAYS

_____

_____

IDEAS FOR NEXT RIDE

DAY

TIME

ARENA/ LOCATION

LUNGING _____ W   T   C

WARM-UP _____ W   T   C

PEAK RIDE _____ W   T   C

COOL DOWN _____ W   T   C

MOOD

INSTRUCTOR _____

BEFORE RIDE   **DURING RIDE**

TAKEAWAYS

_____

_____

IDEAS FOR NEXT RIDE

DAY	TIME	ARENA/ LOCATION

LUNGING _____ W   T   C _____

WARM-UP _____ W   T   C _____

PEAK RIDE _____ W   T   C _____

COOL DOWN _____ W   T   C _____

MOOD   **INSTRUCTOR** _____

BEFORE RIDE   DURING RIDE

TAKEAWAYS

_____

_____

IDEAS FOR NEXT RIDE

DAY	TIME	ARENA/ LOCATION

LUNGING _____ **W**   **T**   **C** _____

WARM-UP _____ **W**   **T**   **C** _____

PEAK RIDE _____ **W**   **T**   **C** _____

COOL DOWN _____ **W**   **T**   **C** _____

MOOD

INSTRUCTOR _____

BEFORE RIDE   **DURING RIDE**

TAKEAWAYS

_____

_____

IDEAS FOR NEXT RIDE

DAY

TIME

ARENA/ LOCATION

LUNGING _____ W T C _____

WARM-UP _____ W T C _____

PEAK RIDE _____ W T C _____

COOL DOWN _____ W T C _____

MOOD

INSTRUCTOR _____

BEFORE RIDE   DURING RIDE

TAKEAWAYS

_____

_____

IDEAS FOR NEXT RIDE

DAY

TIME

ARENA/ LOCATION

LUNGING _____ W T C

WARM-UP _____ W T C

PEAK RIDE _____ W T C

COOL DOWN _____ W T C

MOOD

INSTRUCTOR _____

BEFORE RIDE   DURING RIDE

TAKEAWAYS

_____

_____

IDEAS FOR NEXT RIDE

DAY	TIME	ARENA/ LOCATION

LUNGING _____ **W**   **T**   **C** _____

WARM-UP _____ **W**   **T**   **C** _____

PEAK RIDE _____ **W**   **T**   **C** _____

COOL DOWN _____ **W**   **T**   **C** _____

MOOD

INSTRUCTOR _____

BEFORE RIDE   **DURING RIDE**

TAKEAWAYS

_____

_____

IDEAS FOR NEXT RIDE

DAY

TIME

ARENA/ LOCATION

LUNGING _____ W T C

WARM-UP _____ W T C

PEAK RIDE _____ W T C

COOL DOWN _____ W T C

MOOD

INSTRUCTOR _____

BEFORE RIDE   **DURING RIDE**

TAKEAWAYS

_____

_____

IDEAS FOR NEXT RIDE

DAY	TIME	ARENA/ LOCATION

LUNGING _____ W T C

WARM-UP _____ W T C

PEAK RIDE _____ W T C

COOL DOWN _____ W T C

MOOD   **INSTRUCTOR** _____

BEFORE RIDE   DURING RIDE

TAKEAWAYS

_____

_____

IDEAS FOR NEXT RIDE

DAY

TIME

ARENA/ LOCATION

LUNGING _____ W T C

WARM-UP _____ W T C

PEAK RIDE _____ W T C

COOL DOWN _____ W T C

MOOD

INSTRUCTOR _____

BEFORE RIDE **DURING RIDE**

TAKEAWAYS

_____

_____

IDEAS FOR NEXT RIDE

DAY

TIME

ARENA/ LOCATION

LUNGING _____ **W**   **T**   **C**

WARM-UP _____ **W**   **T**   **C**

PEAK RIDE _____ **W**   **T**   **C**

COOL DOWN _____ **W**   **T**   **C**

MOOD   **INSTRUCTOR** _____

BEFORE RIDE   **DURING RIDE**

TAKEAWAYS

_____

_____

IDEAS FOR NEXT RIDE

DAY

TIME

ARENA/ LOCATION

LUNGING _____ W T C

WARM-UP _____ W T C

PEAK RIDE _____ W T C

COOL DOWN _____ W T C

MOOD   INSTRUCTOR _____

BEFORE RIDE   DURING RIDE

TAKEAWAYS

_____

_____

IDEAS FOR NEXT RIDE

DAY	TIME	ARENA/ LOCATION

LUNGING _____ **W**   **T**   **C** _____

WARM-UP _____ **W**   **T**   **C** _____

PEAK RIDE _____ **W**   **T**   **C** _____

COOL DOWN _____ **W**   **T**   **C** _____

MOOD   **INSTRUCTOR** _____

BEFORE RIDE   **DURING RIDE**

TAKEAWAYS

_____

_____

IDEAS FOR NEXT RIDE

DAY

TIME

ARENA/ LOCATION

LUNGING _____ W T C

WARM-UP _____ W T C

PEAK RIDE _____ W T C

COOL DOWN _____ W T C

mood

INSTRUCTOR _____

BEFORE RIDE   **DURING RIDE**

TAKEAWAYS

_____

_____

IDEAS FOR NEXT RIDE

DAY	TIME	ARENA/ LOCATION

LUNGING _____ W T C _____

WARM-UP _____ W T C _____

PEAK RIDE _____ W T C _____

COOL DOWN _____ W T C _____

MOOD

INSTRUCTOR _____

BEFORE RIDE   DURING RIDE

TAKEAWAYS

_____

_____

IDEAS FOR NEXT RIDE

DAY

TIME

ARENA/ LOCATION

LUNGING _____ W   T   C

WARM-UP _____ W   T   C

PEAK RIDE _____ W   T   C

COOL DOWN _____ W   T   C

MOOD

INSTRUCTOR _____

BEFORE RIDE   **DURING RIDE**

TAKEAWAYS

_____

_____

IDEAS FOR NEXT RIDE

DAY

TIME

ARENA/ LOCATION

LUNGING _____ W   T   C _____

WARM-UP _____ W   T   C _____

PEAK RIDE _____ W   T   C _____

COOL DOWN _____ W   T   C _____

MOOD

INSTRUCTOR _____

BEFORE RIDE   DURING RIDE

TAKEAWAYS

_____

_____

IDEAS FOR NEXT RIDE

DAY

TIME

ARENA/ LOCATION

LUNGING _____ W   T   C _____

WARM-UP _____ W   T   C _____

PEAK RIDE _____ W   T   C _____

COOL DOWN _____ W   T   C _____

MOOD

INSTRUCTOR _____

BEFORE RIDE   **DURING RIDE**

TAKEAWAYS

_____

_____

IDEAS FOR NEXT RIDE

DAY

TIME

ARENA/ LOCATION

LUNGING _____ W   T   C

WARM-UP _____ W   T   C

PEAK RIDE _____ W   T   C

COOL DOWN _____ W   T   C

MOOD

INSTRUCTOR _____

BEFORE RIDE   **DURING RIDE**

TAKEAWAYS

_____

_____

IDEAS FOR NEXT RIDE

DAY

TIME

ARENA/ LOCATION

LUNGING _____ W T C

WARM-UP _____ W T C

PEAK RIDE _____ W T C

COOL DOWN _____ W T C

MOOD

INSTRUCTOR _____

BEFORE RIDE   DURING RIDE

TAKEAWAYS

_____

_____

IDEAS FOR NEXT RIDE

DAY	TIME	ARENA/ LOCATION

LUNGING _____ W   T   C _____

WARM-UP _____ W   T   C _____

PEAK RIDE _____ W   T   C _____

COOL DOWN _____ W   T   C _____

MOOD

INSTRUCTOR _____

BEFORE RIDE   **DURING RIDE**

TAKEAWAYS

_____

_____

IDEAS FOR NEXT RIDE

DAY	TIME	ARENA/ LOCATION

LUNGING _____ W   T   C

WARM-UP _____ W   T   C

PEAK RIDE _____ W   T   C

COOL DOWN _____ W   T   C

MOOD

INSTRUCTOR _____

BEFORE RIDE   **DURING RIDE**

TAKEAWAYS

_____

_____

IDEAS FOR NEXT RIDE

DAY

TIME

ARENA/ LOCATION

LUNGING _____ W   T   C

WARM-UP _____ W   T   C

PEAK RIDE _____ W   T   C

COOL DOWN _____ W   T   C

MOOD

INSTRUCTOR _____

BEFORE RIDE   DURING RIDE

TAKEAWAYS

_____

_____

IDEAS FOR NEXT RIDE

DAY

TIME

ARENA/ LOCATION

LUNGING_____ W   T   C _____

WARM-UP _____ W   T   C _____

PEAK RIDE _____ W   T   C _____

COOL DOWN _____ W   T   C _____

MOOD   INSTRUCTOR _____

BEFORE RIDE   DURING RIDE

TAKEAWAYS

_____

_____

IDEAS FOR NEXT RIDE

DAY	TIME	ARENA/ LOCATION

LUNGING _____ W   T   C _____

WARM-UP _____ W   T   C _____

PEAK RIDE _____ W   T   C _____

COOL DOWN _____ W   T   C _____

MOOD

INSTRUCTOR _____

BEFORE RIDE   **DURING RIDE**

TAKEAWAYS

_____

_____

IDEAS FOR NEXT RIDE

DAY

TIME

ARENA/ LOCATION

LUNGING _____ W   T   C _____

WARM-UP _____ W   T   C _____

PEAK RIDE _____ W   T   C _____

COOL DOWN _____ W   T   C _____

MOOD   INSTRUCTOR _____

BEFORE RIDE   **DURING RIDE**

TAKEAWAYS

_____

_____

IDEAS FOR NEXT RIDE

DAY	TIME	ARENA/ LOCATION

LUNGING _____ W   T   C _____

WARM-UP _____ W   T   C _____

PEAK RIDE _____ W   T   C _____

COOL DOWN _____ W   T   C _____

MOOD   INSTRUCTOR _____

BEFORE RIDE   DURING RIDE

TAKEAWAYS

_____

_____

IDEAS FOR NEXT RIDE

DAY

TIME

ARENA/ LOCATION

LUNGING _____ W T C

WARM-UP _____ W T C

PEAK RIDE _____ W T C

COOL DOWN _____ W T C

MOOD

INSTRUCTOR _____

BEFORE RIDE   DURING RIDE

TAKEAWAYS

_____

_____

IDEAS FOR NEXT RIDE

DAY

TIME

ARENA/ LOCATION

LUNGING _____ W   T   C

WARM-UP _____ W   T   C

PEAK RIDE _____ W   T   C

COOL DOWN _____ W   T   C

MOOD

INSTRUCTOR _____

BEFORE RIDE   **DURING RIDE**

TAKEAWAYS

_____

_____

IDEAS FOR NEXT RIDE

DAY

TIME

ARENA/ LOCATION

LUNGING _____ W   T   C _____

WARM-UP _____ W   T   C _____

PEAK RIDE _____ W   T   C _____

COOL DOWN _____ W   T   C _____

MOOD

INSTRUCTOR _____

BEFORE RIDE   DURING RIDE

TAKEAWAYS

_____

_____

IDEAS FOR NEXT RIDE

DAY	TIME	ARENA/ LOCATION

LUNGING _____ W   T   C

WARM-UP _____ W   T   C

PEAK RIDE _____ W   T   C

COOL DOWN _____ W   T   C

MOOD

INSTRUCTOR _____

BEFORE RIDE   DURING RIDE

TAKEAWAYS

_____

_____

IDEAS FOR NEXT RIDE

DAY	TIME	ARENA/ LOCATION

LUNGING _____ W T C _____

WARM-UP _____ W T C _____

PEAK RIDE _____ W T C _____

COOL DOWN _____ W T C _____

MOOD

INSTRUCTOR _____

BEFORE RIDE   DURING RIDE

TAKEAWAYS

_____

_____

IDEAS FOR NEXT RIDE

DAY	TIME	ARENA/ LOCATION

LUNGING _____ W   T   C _____

WARM-UP _____ W   T   C _____

PEAK RIDE _____ W   T   C _____

COOL DOWN _____ W   T   C _____

MOOD   INSTRUCTOR _____

BEFORE RIDE   **DURING RIDE**

TAKEAWAYS

_____

_____

IDEAS FOR NEXT RIDE

DAY

TIME

ARENA/ LOCATION

LUNGING _____ W T C

WARM-UP _____ W T C

PEAK RIDE _____ W T C

COOL DOWN _____ W T C

MOOD

INSTRUCTOR _____

BEFORE RIDE   DURING RIDE

TAKEAWAYS

_____

_____

IDEAS FOR NEXT RIDE

DAY

TIME

ARENA/ LOCATION

LUNGING _____ W   T   C _____

WARM-UP _____ W   T   C _____

PEAK RIDE _____ W   T   C _____

COOL DOWN _____ W   T   C _____

MOOD

INSTRUCTOR _____

BEFORE RIDE   DURING RIDE

TAKEAWAYS

_____

_____

IDEAS FOR NEXT RIDE

DAY

TIME

ARENA/ LOCATION

LUNGING_____ W   T   C _____

WARM-UP_____ W   T   C _____

PEAK RIDE_____ W   T   C _____

COOL DOWN_____ W   T   C _____

MOOD

INSTRUCTOR _____

BEFORE RIDE   **DURING RIDE**

TAKEAWAYS

_____

_____

IDEAS FOR NEXT RIDE

DAY	TIME	ARENA/ LOCATION

LUNGING _____ W T C _____

WARM-UP _____ W T C _____

PEAK RIDE _____ W T C _____

COOL DOWN _____ W T C _____

MOOD

INSTRUCTOR _____

BEFORE RIDE   DURING RIDE

TAKEAWAYS

_____

_____

_____

IDEAS FOR NEXT RIDE

DAY	TIME	ARENA/ LOCATION

LUNGING _____ W   T   C _____

WARM-UP _____ W   T   C _____

PEAK RIDE _____ W   T   C _____

COOL DOWN _____ W   T   C _____

MOOD

INSTRUCTOR _____

BEFORE RIDE   DURING RIDE

TAKEAWAYS

_____

_____

IDEAS FOR NEXT RIDE

DAY

TIME

ARENA/ LOCATION

LUNGING _____ W T C

WARM-UP _____ W T C

PEAK RIDE _____ W T C

COOL DOWN _____ W T C

MOOD

INSTRUCTOR _____

BEFORE RIDE   DURING RIDE

TAKEAWAYS

_____

_____

IDEAS FOR NEXT RIDE

DAY

TIME

ARENA/ LOCATION

LUNGING _____ W   T   C _____

WARM-UP _____ W   T   C _____

PEAK RIDE _____ W   T   C _____

COOL DOWN _____ W   T   C _____

MOOD

INSTRUCTOR _____

BEFORE RIDE   **DURING RIDE**

TAKEAWAYS

_____
_____
_____

IDEAS FOR NEXT RIDE

DAY	TIME	ARENA/ LOCATION

LUNGING _____ W   T   C _____

WARM-UP _____ W   T   C _____

PEAK RIDE _____ W   T   C _____

COOL DOWN _____ W   T   C _____

MOOD   INSTRUCTOR _____

BEFORE RIDE   DURING RIDE

TAKEAWAYS

_____

_____

IDEAS FOR NEXT RIDE

DAY

TIME

ARENA/ LOCATION

LUNGING _____ W   T   C

WARM-UP _____ W   T   C

PEAK RIDE _____ W   T   C

COOL DOWN _____ W   T   C

MOOD

INSTRUCTOR _____

BEFORE RIDE   **DURING RIDE**

TAKEAWAYS

_____

_____

IDEAS FOR NEXT RIDE

DAY	TIME	ARENA/ LOCATION

LUNGING _____ **W**   **T**   **C** _____

WARM-UP _____ **W**   **T**   **C** _____

PEAK RIDE _____ **W**   **T**   **C** _____

COOL DOWN _____ **W**   **T**   **C** _____

MOOD

INSTRUCTOR _____

BEFORE RIDE   **DURING RIDE**

TAKEAWAYS

_____

_____

_____

IDEAS FOR NEXT RIDE

DAY

TIME

ARENA/ LOCATION

LUNGING _____ W T C

WARM-UP _____ W T C

PEAK RIDE _____ W T C

COOL DOWN _____ W T C

MOOD

INSTRUCTOR _____

BEFORE RIDE   **DURING RIDE**

TAKEAWAYS

_____

_____

_____

IDEAS FOR NEXT RIDE

DAY

TIME

ARENA/ LOCATION

LUNGING _____ W   T   C

WARM-UP _____ W   T   C

PEAK RIDE _____ W   T   C

COOL DOWN _____ W   T   C

MOOD   **INSTRUCTOR** _____

BEFORE RIDE   **DURING RIDE**

TAKEAWAYS

_____

_____

IDEAS FOR NEXT RIDE

DAY

TIME

ARENA/ LOCATION

LUNGING _____ W T C

WARM-UP _____ W T C

PEAK RIDE _____ W T C

COOL DOWN _____ W T C

MOOD

INSTRUCTOR _____

BEFORE RIDE   **DURING RIDE**

TAKEAWAYS

_____

_____

IDEAS FOR NEXT RIDE

DAY

TIME

ARENA/ LOCATION

LUNGING _____ W   T   C

WARM-UP _____ W   T   C

PEAK RIDE _____ W   T   C

COOL DOWN _____ W   T   C

MOOD

INSTRUCTOR _____

BEFORE RIDE   **DURING RIDE**

TAKEAWAYS

_____

_____

IDEAS FOR NEXT RIDE

DAY

TIME

ARENA/ LOCATION

LUNGING _____ W   T   C _____

WARM-UP _____ W   T   C _____

PEAK RIDE _____ W   T   C _____

COOL DOWN _____ W   T   C _____

MOOD

INSTRUCTOR _____

BEFORE RIDE   **DURING RIDE**

TAKEAWAYS

_____

_____

IDEAS FOR NEXT RIDE

DAY	TIME	ARENA/ LOCATION

LUNGING _____ W   T   C _____

WARM-UP _____ W   T   C _____

PEAK RIDE _____ W   T   C _____

COOL DOWN _____ W   T   C _____

MOOD

INSTRUCTOR _____

BEFORE RIDE   DURING RIDE

TAKEAWAYS

_____

_____

IDEAS FOR NEXT RIDE

DAY	TIME	ARENA/ LOCATION

LUNGING _____ W   T   C

WARM-UP _____ W   T   C

PEAK RIDE _____ W   T   C

COOL DOWN _____ W   T   C

MOOD

INSTRUCTOR _____

BEFORE RIDE   DURING RIDE

TAKEAWAYS

_____

_____

IDEAS FOR NEXT RIDE

DAY

TIME

ARENA/ LOCATION

LUNGING_____ **W**   **T**   **C** _____

WARM-UP _____ **W**   **T**   **C** _____

PEAK RIDE _____ **W**   **T**   **C** _____

COOL DOWN _____ **W**   **T**   **C** _____

MOOD   **INSTRUCTOR** _____

BEFORE RIDE   **DURING RIDE**

TAKEAWAYS

_____

_____

IDEAS FOR NEXT RIDE

DAY

TIME

ARENA/ LOCATION

LUNGING _____ W T C

WARM-UP _____ W T C

PEAK RIDE _____ W T C

COOL DOWN _____ W T C

mood

INSTRUCTOR _____

BEFORE RIDE   **DURING RIDE**

TAKEAWAYS

_____

_____

IDEAS FOR NEXT RIDE

DAY

TIME

ARENA/ LOCATION

LUNGING _____ W T C

WARM-UP _____ W T C

PEAK RIDE _____ W T C

COOL DOWN _____ W T C

MOOD

INSTRUCTOR _____

BEFORE RIDE   DURING RIDE

TAKEAWAYS

_____

_____

IDEAS FOR NEXT RIDE

DAY

TIME

ARENA/ LOCATION

LUNGING _____ W   T   C _____

WARM-UP _____ W   T   C _____

PEAK RIDE _____ W   T   C _____

COOL DOWN _____ W   T   C _____

MOOD

INSTRUCTOR _____

BEFORE RIDE   DURING RIDE

TAKEAWAYS

_____

_____

IDEAS FOR NEXT RIDE

DAY	**TIME**	**ARENA/ LOCATION**

LUNGING _____ **W**   **T**   **C** _____

WARM-UP _____ **W**   **T**   **C** _____

PEAK RIDE _____ **W**   **T**   **C** _____

COOL DOWN _____ **W**   **T**   **C** _____

MOOD   **INSTRUCTOR** _____

BEFORE RIDE   **DURING RIDE**

TAKEAWAYS

_____

_____

IDEAS FOR NEXT RIDE

DAY	TIME	ARENA/ LOCATION

LUNGING _____ W   T   C _____

WARM-UP _____ W   T   C _____

PEAK RIDE _____ W   T   C _____

COOL DOWN _____ W   T   C _____

MOOD

INSTRUCTOR _____

BEFORE RIDE   **DURING RIDE**

TAKEAWAYS

_____

_____

IDEAS FOR NEXT RIDE

DAY

TIME

ARENA/ LOCATION

LUNGING _____ W   T   C _____

WARM-UP _____ W   T   C _____

PEAK RIDE _____ W   T   C _____

COOL DOWN _____ W   T   C _____

MOOD

INSTRUCTOR _____

BEFORE RIDE   DURING RIDE

TAKEAWAYS

_____

_____

IDEAS FOR NEXT RIDE

DAY

TIME

ARENA/ LOCATION

LUNGING _____ W T C

WARM-UP _____ W T C

PEAK RIDE _____ W T C

COOL DOWN _____ W T C

MOOD

INSTRUCTOR _____

BEFORE RIDE   **DURING RIDE**

TAKEAWAYS

_____

_____

IDEAS FOR NEXT RIDE

DAY

TIME

ARENA/ LOCATION

LUNGING _____ W T C

WARM-UP _____ W T C

PEAK RIDE _____ W T C

COOL DOWN _____ W T C

MOOD

INSTRUCTOR _____

BEFORE RIDE   **DURING RIDE**

TAKEAWAYS

_____

_____

IDEAS FOR NEXT RIDE

DAY	TIME	ARENA/ LOCATION

LUNGING _____ **W**   **T**   **C** _____

WARM-UP _____ **W**   **T**   **C** _____

PEAK RIDE _____ **W**   **T**   **C** _____

COOL DOWN _____ **W**   **T**   **C** _____

MOOD

INSTRUCTOR _____

BEFORE RIDE   **DURING RIDE**

TAKEAWAYS

_____

_____

IDEAS FOR NEXT RIDE

DAY	**TIME**	**ARENA/ LOCATION**

LUNGING _____ W   T   C _____

WARM-UP _____ W   T   C _____

PEAK RIDE _____ W   T   C _____

COOL DOWN _____ W   T   C _____

MOOD

INSTRUCTOR _____

BEFORE RIDE   **DURING RIDE**

TAKEAWAYS

_____

_____

IDEAS FOR NEXT RIDE

DAY	TIME	ARENA/ LOCATION

LUNGING _____ W   T   C

WARM-UP _____ W   T   C

PEAK RIDE _____ W   T   C

COOL DOWN _____ W   T   C

MOOD

INSTRUCTOR _____

BEFORE RIDE   **DURING RIDE**

TAKEAWAYS

_____

_____

IDEAS FOR NEXT RIDE

DAY

TIME

ARENA/ LOCATION

LUNGING _____ W   T   C

WARM-UP _____ W   T   C

PEAK RIDE _____ W   T   C

COOL DOWN _____ W   T   C

MOOD

INSTRUCTOR _____

BEFORE RIDE   **DURING RIDE**

TAKEAWAYS

_____

_____

IDEAS FOR NEXT RIDE

DAY

TIME

ARENA/ LOCATION

LUNGING _____ W   T   C

WARM-UP _____ W   T   C

PEAK RIDE _____ W   T   C

COOL DOWN _____ W   T   C

MOOD

INSTRUCTOR _____

BEFORE RIDE   **DURING RIDE**

TAKEAWAYS

_____

_____

IDEAS FOR NEXT RIDE

DAY

TIME

ARENA/ LOCATION

LUNGING _____ W   T   C

WARM-UP _____ W   T   C

PEAK RIDE _____ W   T   C

COOL DOWN _____ W   T   C

MOOD

INSTRUCTOR _____

BEFORE RIDE   **DURING RIDE**

TAKEAWAYS

_____

_____

IDEAS FOR NEXT RIDE

DAY

TIME

ARENA/ LOCATION

LUNGING_____ W   T   C

WARM-UP _____ W   T   C

PEAK RIDE _____ W   T   C

COOL DOWN_____ W   T   C

MOOD   INSTRUCTOR _____

BEFORE RIDE   DURING RIDE

TAKEAWAYS

_____

_____

IDEAS FOR NEXT RIDE

DAY

TIME

ARENA/ LOCATION

LUNGING _____ W   T   C

WARM-UP _____ W   T   C

PEAK RIDE _____ W   T   C

COOL DOWN _____ W   T   C

MOOD

INSTRUCTOR _____

BEFORE RIDE   DURING RIDE

TAKEAWAYS

_____

_____

IDEAS FOR NEXT RIDE

DAY	TIME	ARENA/ LOCATION

LUNGING _____ W   T   C

WARM-UP _____ W   T   C

PEAK RIDE _____ W   T   C

COOL DOWN _____ W   T   C

MOOD   INSTRUCTOR _____

BEFORE RIDE   DURING RIDE

TAKEAWAYS

_____

_____

IDEAS FOR NEXT RIDE

DAY

TIME

ARENA/ LOCATION

LUNGING _____ W T C

WARM-UP _____ W T C

PEAK RIDE _____ W T C

COOL DOWN _____ W T C

MOOD

INSTRUCTOR _____

BEFORE RIDE   **DURING RIDE**

TAKEAWAYS

_____

_____

IDEAS FOR NEXT RIDE

DAY

TIME

ARENA/ LOCATION

LUNGING _____ W   T   C _____

WARM-UP _____ W   T   C _____

PEAK RIDE _____ W   T   C _____

COOL DOWN _____ W   T   C _____

MOOD   **INSTRUCTOR** _____

BEFORE RIDE   **DURING RIDE**

TAKEAWAYS

_____

_____

IDEAS FOR NEXT RIDE

DAY

TIME

ARENA/ LOCATION

LUNGING _____ W   T   C

WARM-UP _____ W   T   C

PEAK RIDE _____ W   T   C

COOL DOWN _____ W   T   C

MOOD

INSTRUCTOR _____

BEFORE RIDE   DURING RIDE

TAKEAWAYS

_____

_____

IDEAS FOR NEXT RIDE

QUICK REFERENCES

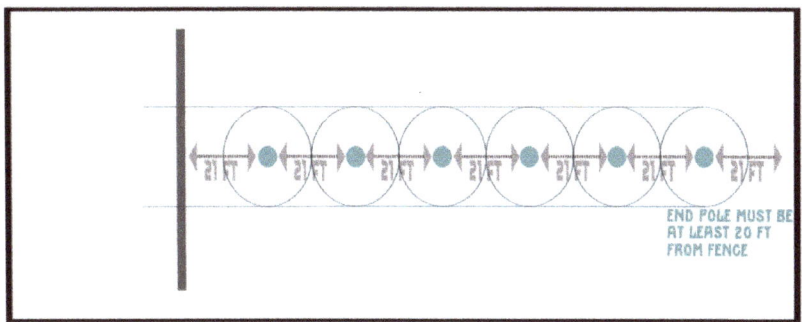

END POLE MUST BE
AT LEAST 20 FT
FROM FENCE

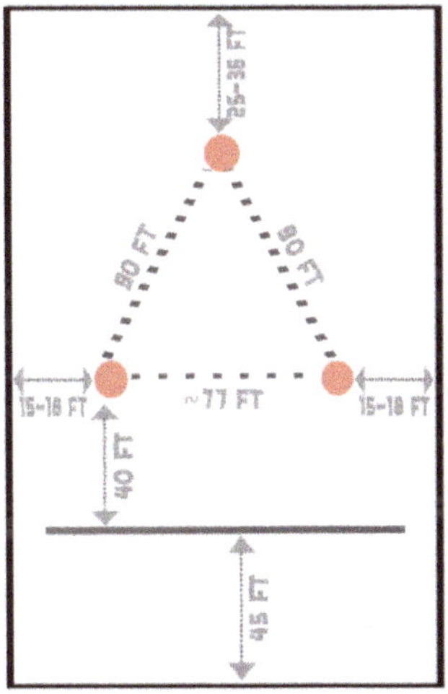

21 FT 21 FT 21 FT 21 FT 21 FT 21 FT 21 FT

SMALL ARENA

25-35 FT

80 FT   80 FT

15-18 FT   ~ 77 FT   15-18 FT

40 FT

45 FT

THE SCORELINE SHOULD BE AT LEAST 45 FT AWAY FROM
THE END OF THE ARENA UNLESS THERE IS A CENTER
ALLEY OR SIZE DOES NOT PERMIT.

BARREL 3 SHOULD BE AT LEAST 15' LONGER THAN THE
BARRELS 1 AND 2.

BARRELS 1 AND 2 SHOULD BE AT LEAST 60 FT APART.

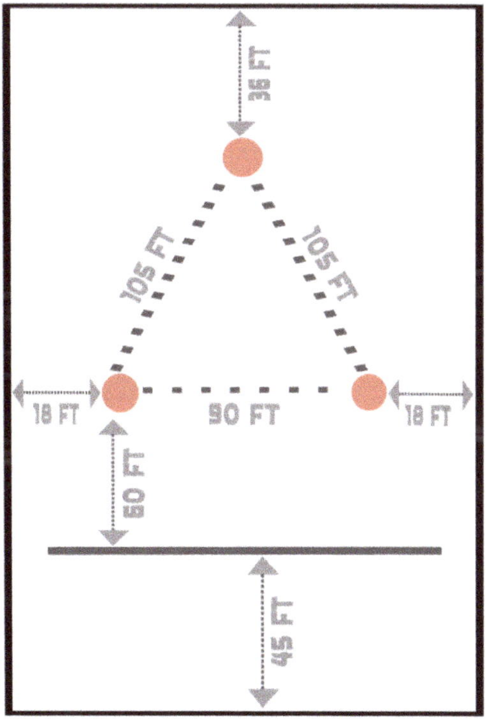

STANDARD ARENA

35 FT

105 FT   105 FT

18 FT   90 FT   18 FT

60 FT

45 FT

QUICK REFERENCES

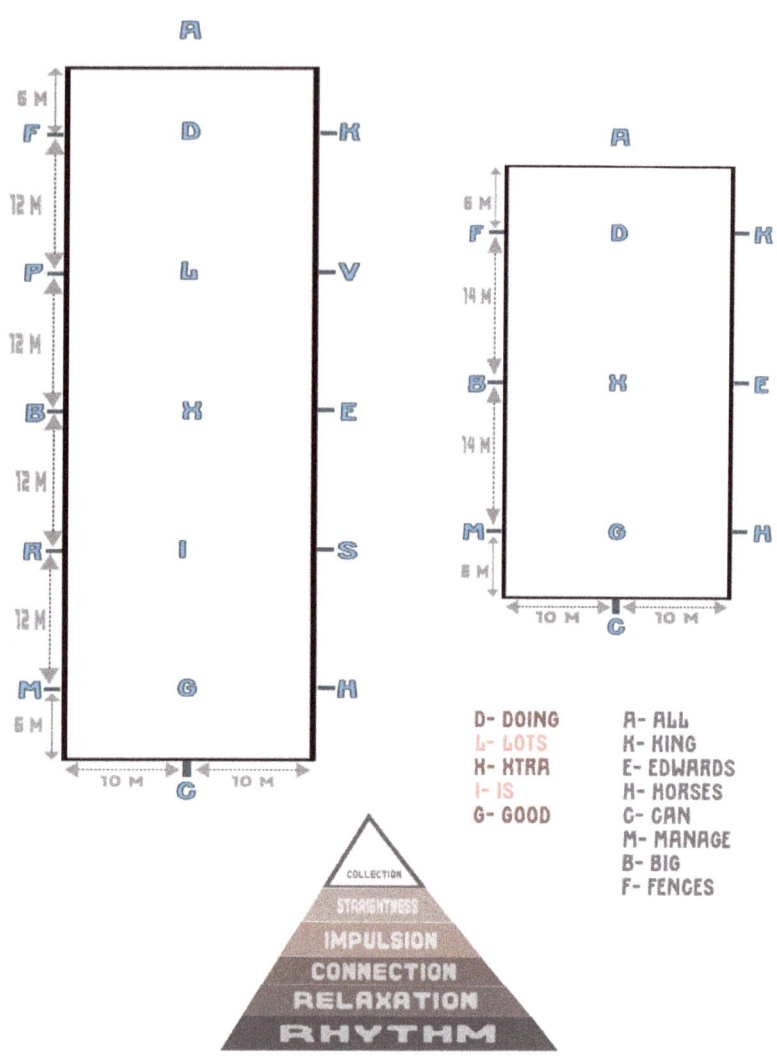

D- DOING
L- LOTS
X- XTRA
I- IS
G- GOOD

A- ALL
K- KING
E- EDWARDS
H- HORSES
C- CAN
M- MANAGE
B- BIG
F- FENCES

COLLECTION
STRAIGHTNESS
IMPULSION
CONNECTION
RELAXATION
RHYTHM

Stevie Lou
DÍAZ

WWW.STEVIELOUDIAZ.COM

SPEEDY PONY PRESS